discovermore
Space Science

Exploring the Sun

Elizabeth Pruet

Britannica Educational Publishing
IN ASSOCIATION WITH
THE ROSEN PUBLISHING GROUP

Published in 2026 by Britannica Educational Publishing (a trademark of Encyclopædia Britannica, Inc.) in association with The Rosen Publishing Group, Inc.
2544 Clinton Street, Buffalo, NY 14224

Distributed exclusively by Rosen Publishing.
To see additional Britannica Educational Publishing titles, go to rosenpublishing.com.

Editor: Greg Roza
Book Design: Michael Flynn

Photo Credits: Cover rawinyeepinyo/Shutterstock.com; (series background) Dai Yim/Shutterstock.com; p. 4 lovelyday12/Shutterstock.com; pp. 5, 8, 13 (all), 14, 17 courtesy of NASA; p. 6 BlueRingMedia/Shutterstock.com; p. 7 Nazarii_Neshcherenskyi/Shutterstock.com; p. 9 Nebojsa S/Shutterstock.com; p. 10 Lars Poyansky/Shutterstock.com; p. 11 Natalllenka.m/Shutterstock.com; p. 15 Abstract51/Shutterstock.com; p. 16 Jemastock/Shutterstock.com; p. 18 Radowitz/Shutterstock.com; p. 19 Ozant/Shutterstock.com; p. 20 GraphicsRF.com/Shutterstock.com; p. 21 Viktor Gabyshev/Shutterstock.com; pp. 22, 23 Peter Hermes Furian/Shutterstock.com; p. 24 Vladfotograf/Shutterstock.com; p. 25 https://commons.wikimedia.org/wiki/File:Parker_Solar_Probe_coronal_stream_wispr-big_1-st_flyby.jpg; p. 26 Radzas2008/Shutterstock.com; p. 27 Chris Harwood/Shutterstock.com; p. 28 Designua/Shutterstock.com; p. 29 Elena11/Shutterstock.com.

Library of Congress Cataloging-in-Publication Data

Names: Pruet, Elizabeth author
Title: Exploring the sun / Elizabeth Pruet.
Other titles: Discover more. Space science
Description: [Buffalo] : Britannica Educational Publishing, [2026] | Series: Discover more: space science | Includes bibliographical references and index.
Identifiers: LCCN 2025038963 | ISBN 9781641905367 library binding | ISBN 9781641905350 paperback | ISBN 9781641905374 ebook
Subjects: LCSH: Sun--Juvenile literature
Classification: LCC QB521.5 .P78 2026
LC record available at https://lccn.loc.gov/2025038963

Manufactured in the United States of America

CPSIA Compliance Information: Batch #CWBRIT26. For further information contact Rosen Publishing at 1-800-237-9932.

Contents

Sun Basics

The sun warms Earth and lights our days, but what is a sun? A sun is a star that is the center of a solar system. All suns are stars, but not all stars are suns because some stars exist without planets surrounding them.

Our sun gives us light and heat energy, helping plants to grow, which then feed us and other animals.

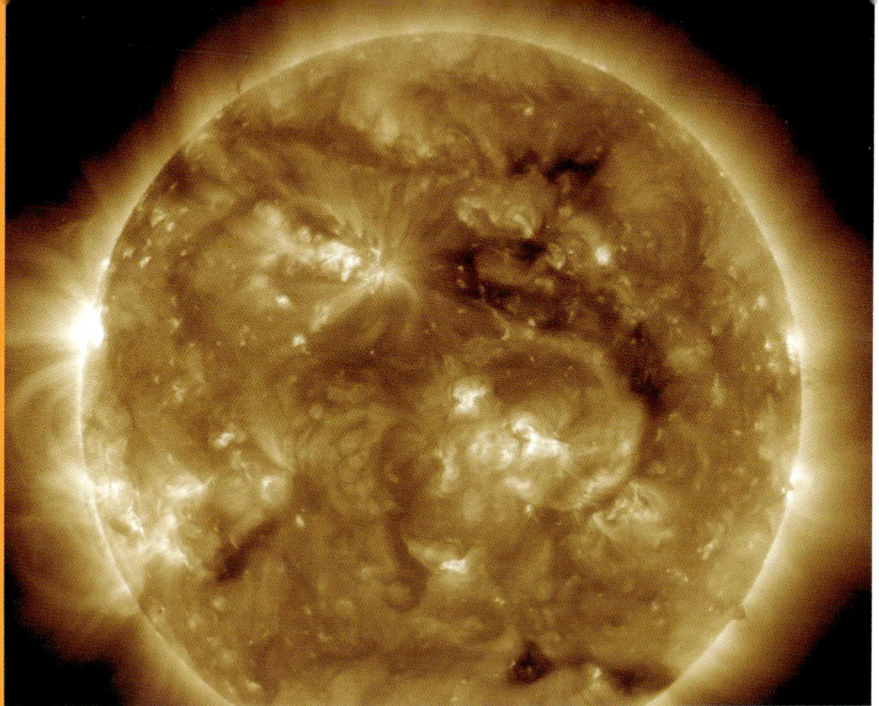

From Earth, the sun looks like a white, glowing ball in the sky. Using telescopes, though, we see that it's a spinning mass of gas called plasma.

A star is a hot sphere, or ball, of gases that gives off energy in the form of light and heat. Our sun is one of many stars in the universe. Scientists think there are at least 100 billion galaxies in the universe, possibly up to 2 trillion. In our galaxy alone (the Milky Way), there are between 100 billion and 400 billion stars. How many stars must there be if each of those galaxies had as many stars as the Milky Way?

Consider This

Most solar systems have two or more stars. What would a sunrise be like with two suns?

Star Classification

There are many types of stars in the universe. They range in color from blue to yellow to red. Blue stars are hotter than red stars, and yellow stars are in the midrange. Hotter stars tend to be brighter than cooler stars. However, the size of a star also affects the brightness.

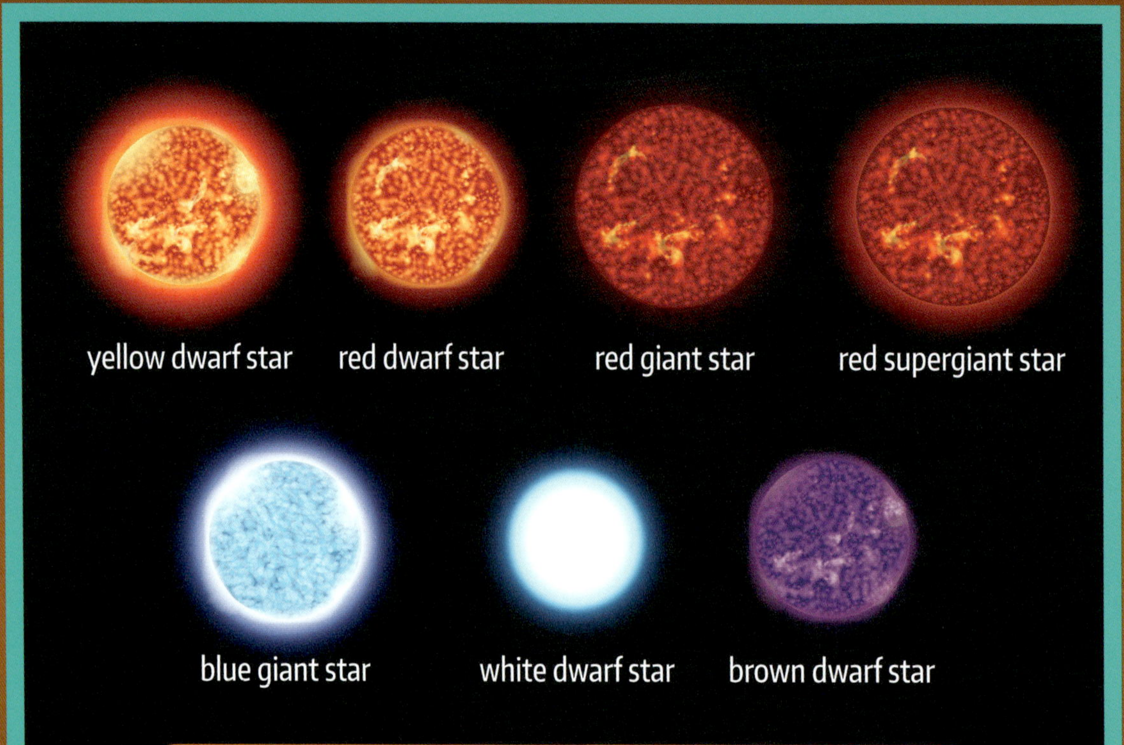

Our sun is a yellow dwarf star, though its light is white.

The sun's closest star neighbor is Proxima Centauri, a red dwarf. This picture shows how much smaller it is compared to the sun.

Red giants and supergiant stars are cooler than other stars but bright because of their size. (The width of a supergiant can be 100 times that of our sun.) In comparison, a white dwarf is hot but dim because it is small. Our sun, a yellow dwarf, is in the midrange of size and brightness. Scientists use the Morgan-Keenan (MK) system to classify each star.

Consider This

HP 56948, a star 200 light-years away from Earth, has the same mass, temperature, and chemical makeup as our sun. Scientists wonder if it has a planet in its solar system like Earth. What might this other Earth be like?

Neighboring Stars

Our neighbor, Proxima Centauri, is part of a three-star system known as Alpha Centauri. *Proxima* means "nearest" in Latin. Alpha Centauri A is a yellow star. Alpha Centauri B is redder. They orbit each other and together are the fourth-brightest star in Earth's sky. Proxima Centauri orbits the other two stars and is so faint we can't see it with the naked eye.

These stars are not thought to be orbited by planets that support life.

Proxima Centauri is so faint and far away from the two main stars in Alpha Centauri, it's often not seen in pictures such as this.

A star that has the same mass, temperature, and chemical makeup as the sun is HIP 56948. It is 208 light-years away. Scientists wonder if it has a planet in its solar system like Earth. If so, it might be home to similar life-forms. You can find HIP 56948 in the constellation Draco.

Consider This

When you see a star that is 10 light-years away, you are looking 10 years into the past.

Inside the Sun

Let's look more closely at our sun. Its core, or center, consists mostly of the gases hydrogen and helium. Its core is very hot, probably reaching about 28,080,000°F (15,600,000°C). Hydrogen is packed in tightly at the core, creating a **density** 150 times greater than water.

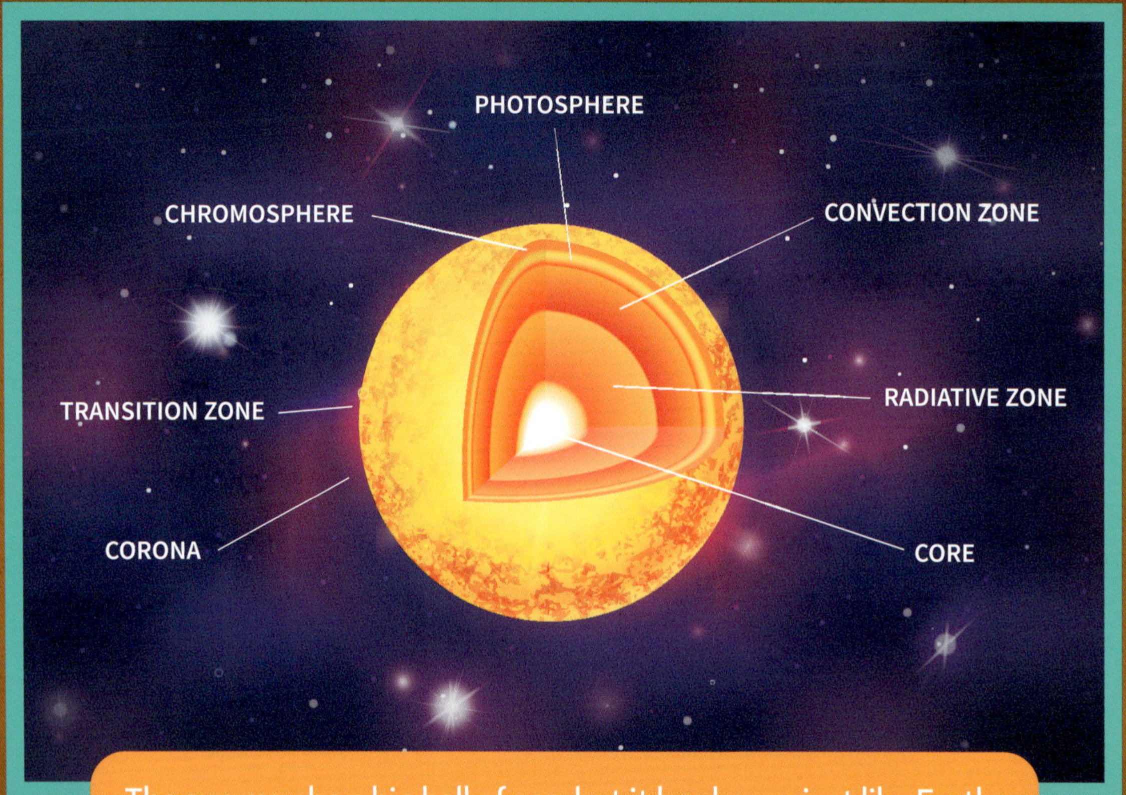

The sun may be a big ball of gas, but it has layers just like Earth.

If you were able to stand on different planets or stars, your weight would be different based on the pressure. What would your weight be on the sun?

Have you ever dived deep into a swimming pool and felt your ears hurt? That was because of the water pressure. The sun's pressure is much greater than that! The pressure causes hydrogen atoms to fuse, or connect, with each other, changing into helium. This gives off huge amounts of energy: the sun's heat and light. This energy then travels to Earth, where it warms our planet and grows our food.

WORD WISE

DENSITY IS THE QUANTITY OF SOMETHING, SUCH AS MOLECULES, PER UNIT VOLUME, UNIT AREA, OR UNIT LENGTH.

The Surface

The outermost layer, or surface, of our sun is called the photosphere. In ancient Greek, *photos* means "light," and *sphere* refers to the sun's ball-like shape. The energy created in the sun's core spreads out through its photosphere, which also consists of gas. This means that even if the sun were not so hot, you could not stand on it. The sun's surface is, of course, very hot, though not as hot as its core. The surface temperature is about 10,000° F (6,000° C).

The surface lets off most of the light and heat that reach Earth. Sometimes cooler, darker patches called sunspots appear on the surface. They appear and disappear in 11-year cycles.

Like Earth, the sun rotates. Because its surface is made of gas, the entire sun does not rotate at the same speed. Its equator rotates every 25 days. At the poles, it rotates every 35 days. The sun's axis tilts about 7.25 degrees, much less than Earth's, which is 23.5 degrees.

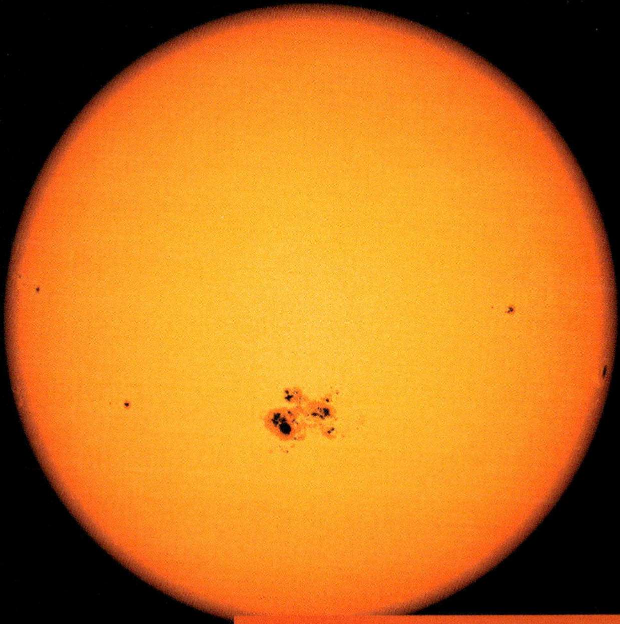

Scientists were able to use the movement of the sun's sunspots to learn how fast it rotates or spins.

compare and contrast

How does Earth's surface differ from the sun's surface?

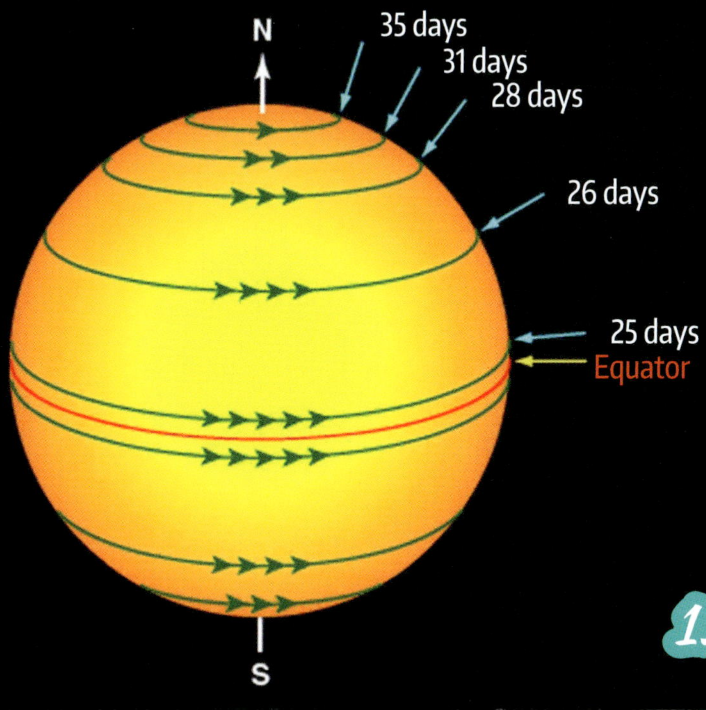

This picture shows how long it takes each part of the sun to complete a full rotation.

The Atmosphere

There are layers to the surface of the sun in an area called the **atmosphere**. The photosphere is the first layer of the sun's atmosphere. The next layer is called the chromosphere. It is visible as a thin, reddish ring around the edge of the sun during total solar eclipses.

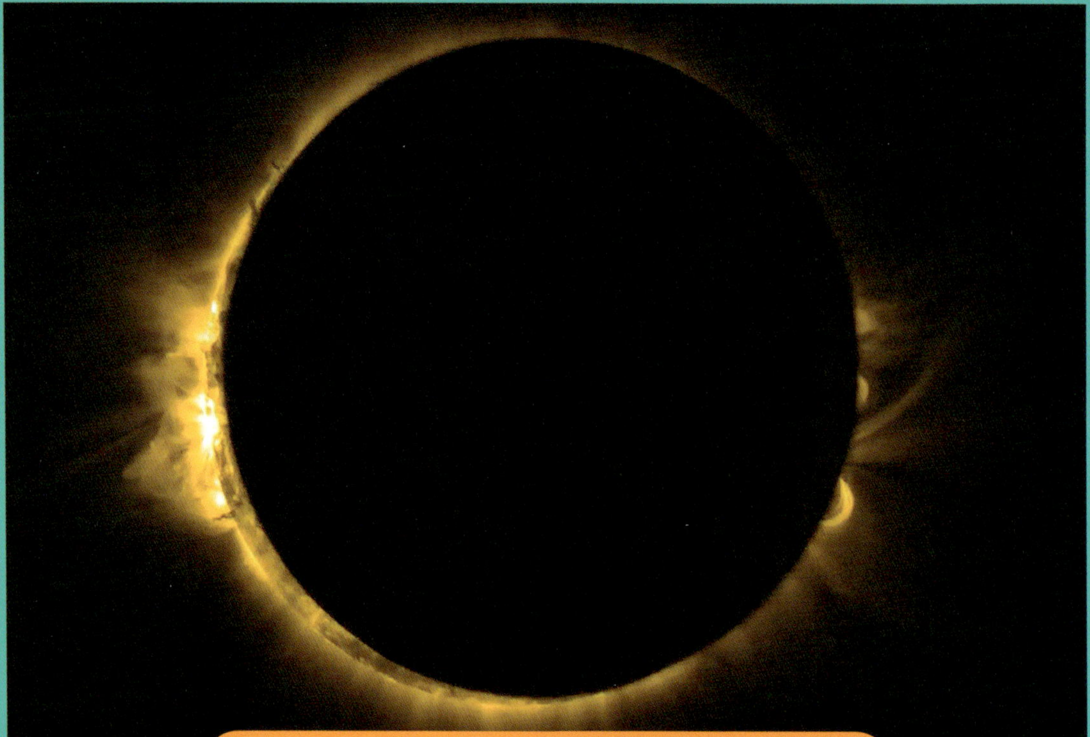

The sun's swirling corona can be seen in this picture during a near-total eclipse in 2015.

The auroras can usually only be seen near the Earth's North and South Poles.

The outer atmosphere is called the corona. It is visible as a white halo during a solar eclipse. The sun's hot corona (reaching 3,600,000 °F or 2,000,000 °C) shoots out particles that carry an electric charge. The stream of particles is called the solar wind. The wind moves out through the solar system at about 250 miles (400 km) per second. On Earth, these solar winds can cause colorful bands of light called auroras to appear in the far northern and southern skies. The Earth's magnetic field protects the planet from most of the harmful solar wind.

WORD WISE

AN ATMOSPHERE IS A LAYER OF GASES SURROUNDING A HEAVENLY BODY, SUCH AS A PLANET OR SUN.

The Solar System

Some stars have planets that orbit them, called planetary systems. Ours is called the solar system because the word "solar" relates to the sun.

Planets orbit a sun because of its strong gravitational pull. The more mass an object has, the greater its gravitational pull. Our sun contains more than 99 percent of the mass contained in the entire solar system.

It takes each planet in our system a different amount of time to completely orbit the sun. For Earth, it's about 365 days.

Jupiter, the largest planet in the solar system, also experiences auroras from the solar wind, as seen at the top of the image.

The four planets closest to the sun have solid surfaces. Mercury is closest, orbiting at a distance of 36 million miles (58 million km). For this reason, temperatures on Mercury can be higher than 800°F (430°C) during the day.

The four outer planets consist of liquid and gas. The largest of these "gas giants" is Jupiter. Like the sun, it is made up of hydrogen and helium. If Jupiter were 80 times larger, it would be massive enough to create the energy needed to be a star. As it is, Jupiter lies at the center of dozens of orbiting moons. Our solar system also has five confirmed dwarf planets: Ceres, Eris, Haumea, Makemake, and Pluto.

Consider This

Some "stars" in the night sky are not stars at all but the planets Mercury, Mars, Venus, Saturn, and Jupiter. From some of these same planets, Earth appears like a star.

Warming the Earth

Earth gets nearly all its energy and heat from the sun. If Earth relied on the sun's heat alone, the average temperature would be 0°F (-18°C), too cold for most living things.

In reality, Earth is 58°F (14°C) on average. This is because of the greenhouse effect. When the sun's rays hit Earth, some heat is absorbed, or taken in. But much of it is bounced back out into space. Some of this heat is trapped by gases in Earth's atmosphere. This also means that Earth stays warm at night.

The term "greenhouse" comes from buildings people use to grow plants. The walls and ceilings are glass or clear plastic, letting in light but also trapping in heat.

Solar radiation, a type of energy, travels from the sun to Earth. Some of it is trapped in Earth's atmosphere, heating up the planet.

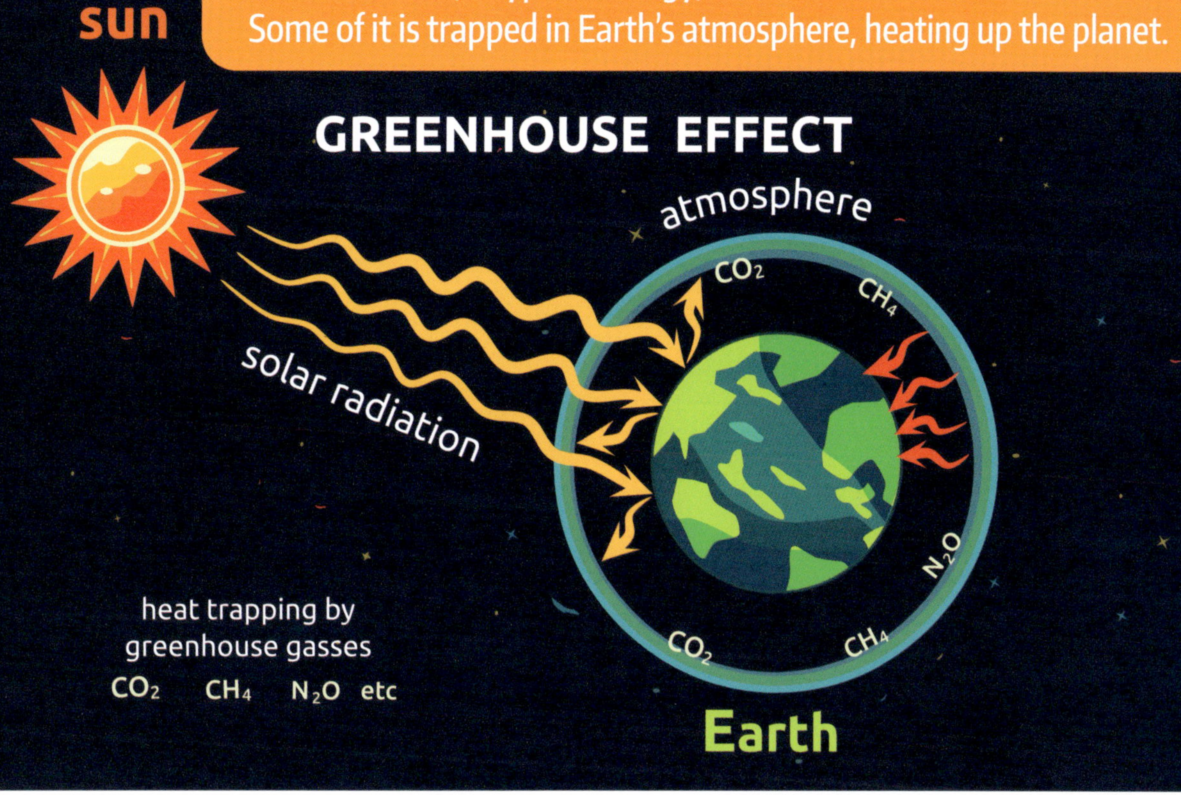

The more chemicals there are in the atmosphere, such as carbon dioxide (CO_2), methane (CH_4), and nitrous oxide (N_2O), the warmer Earth becomes. These chemicals are added to the atmosphere by pollution and are a major source of climate change.

Consider This

Though broiling hot during the day, Mercury's temperature drops to – 300°F (– 180°C) at night. What does Mercury lack that Earth has?

The Angle of the Sun's Rays

Because of the Earth's tilt, the sun doesn't always hit the planet in the same way. The sun's rays hit Earth directly at the **equator** during two seasons of the year. That means the rays are directed at a small area. Where the rays enter at more of an angle, toward the North and South Poles, the heat from the sun is spread out over a larger area. Imagine a space heater used to heat one room versus a whole house. A room would get very warm, but a whole house would remain pretty cold. The same is true at the equator, where it is always warm, and toward the poles, where it is cold.

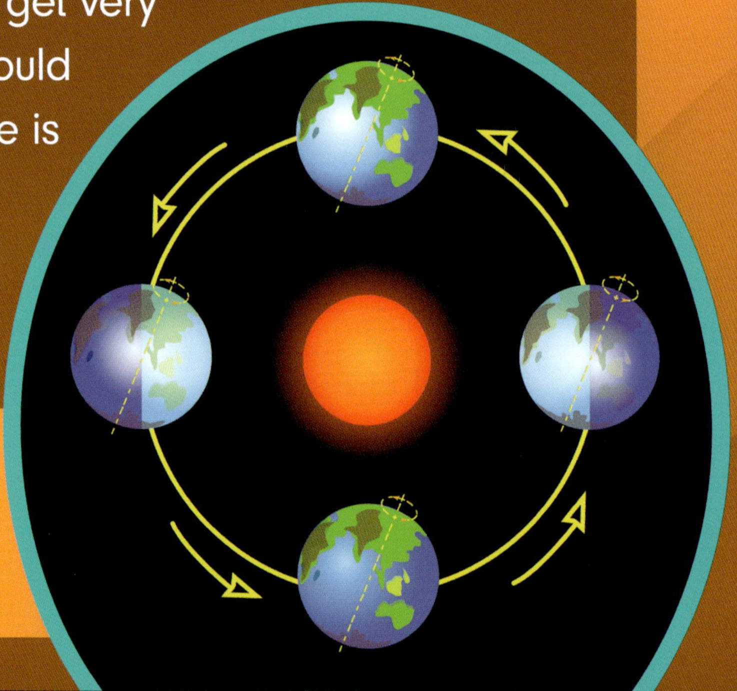

This drawing shows how the sun's light hits the Earth differently throughout the Earth because of the planet's tilt.

This timelapse photo shows how the sun never fully sets near the North Pole during the summer.

Because of Earth's tilt, hours of daylight can be very different at the poles. During the summer, the Arctic has 24 hours of sunlight. This is called the midnight sun. During the winter, it has round-the-clock darkness, a time called polar night. The same is true of Antarctica, only at opposite times of the year.

WORD WISE

THE EQUATOR IS AN IMAGINARY CIRCLE AROUND EARTH THAT IS EQUALLY DISTANT FROM THE NORTH POLE AND THE SOUTH POLE.

The Seasons

The Earth's four seasons happen entirely because of the planet's tilt. In June, July, and August, the Northern Hemisphere is tipped slightly toward the sun. The sun hits as far north as the Tropic of Cancer, creating summer in the Northern Hemisphere. June 21 or 22 marks the summer solstice, when the sun is the farthest north.

summer solstice (June 21)

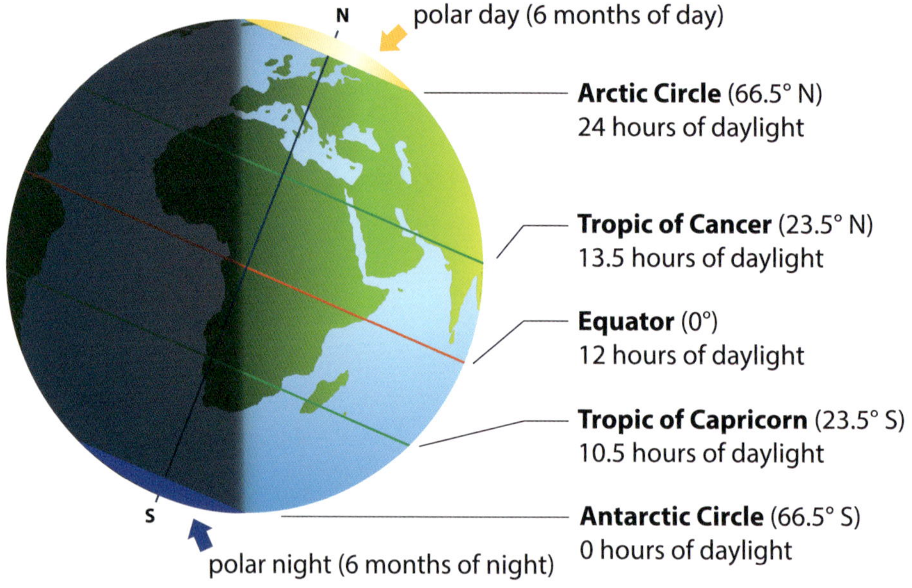

This picture shows how the sun's light hits Earth on June 21 each year.

winter solstice (December 21)

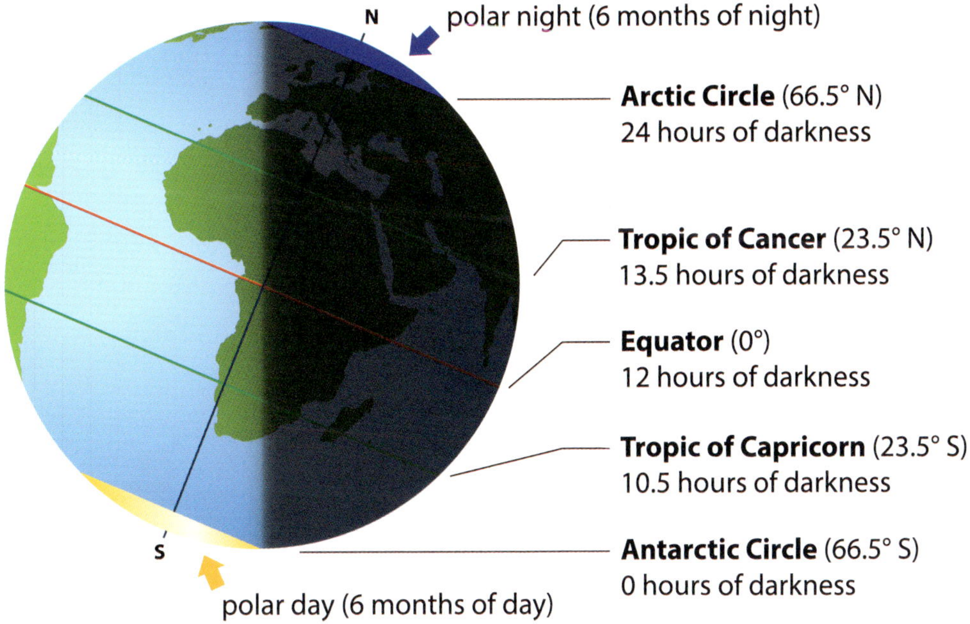

This is how the sun's light hits Earth on December 21.

Winter solstice, when the sun is farthest south, is on December 21 or 22.

In December, the sun hits as far south as the Tropic of Capricorn, causing summer in the Southern Hemisphere. The summer solstice here is on December 21 or 22, and the winter solstice is on June 21 or 22. Imagine what the seasons would be like if the Earth had no tilt.

Consider This

When it is summer in the Northern Hemisphere, it is winter in the Southern Hemisphere, and vice versa. Why does it stay warm near the equator year-round?

Understanding the Sun

Human understanding of the sun has changed over history. In England, between 3100 and 1500 BCE, builders arranged some of the large standing stones at Stonehenge to frame the sunrise during the summer solstice. At around the same time, ancient Egyptians believed in Ra, the sun god.

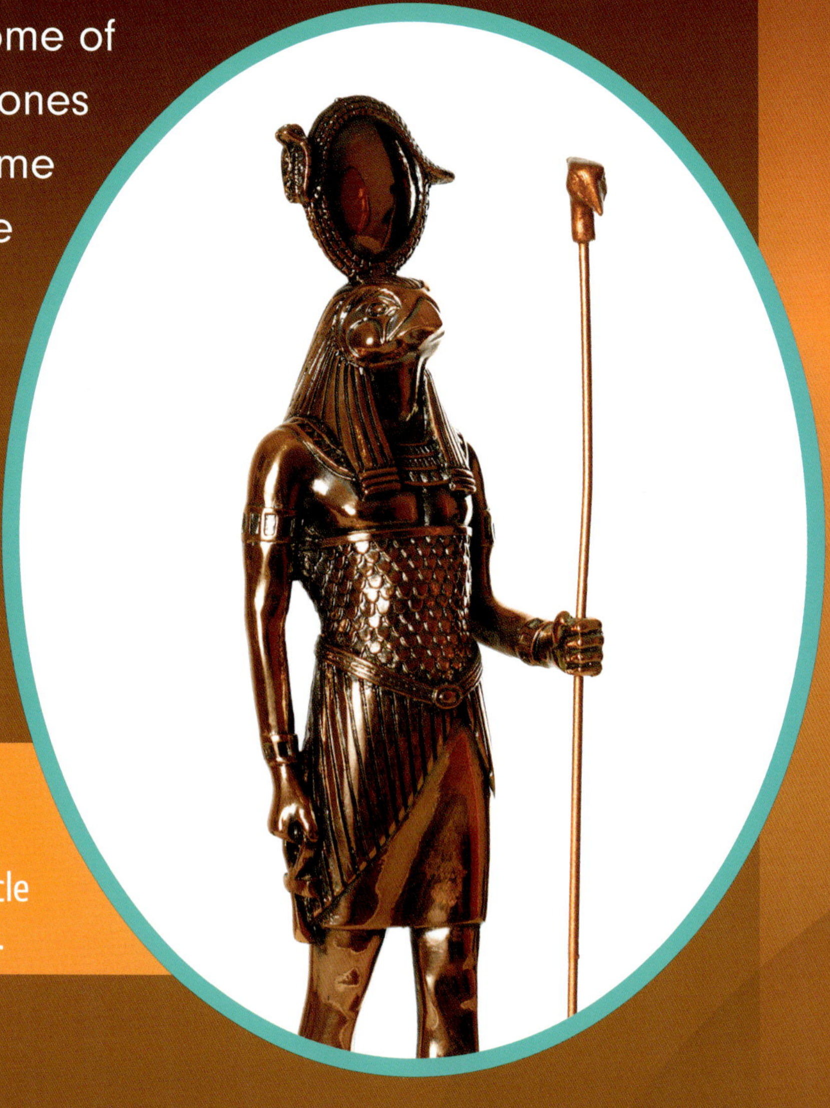

The Egyptian sun god Ra is often shown with the head of a falcon with a circle representing the sun on it.

No other human-made object has come as close to the sun as NASA's Parker Solar Probe. It took this photo of a coronal streamer nearly 17 million miles from the sun's surface.

Later, scientists found ways to study the sun. Nicolaus Copernicus argued (in 1543) that Earth orbits the sun. And in 1686, Isaac Newton published a book explaining how the sun's gravitational pull causes the planets to orbit it. In modern times, scientists have studied the sun's **spectrum** to learn what chemicals make up the sun. They have also sent satellites into space to photograph the sun and to take measurements.

WORD WISE

A SPECTRUM IS THE GROUP OF COLORS, INCLUDING RED, ORANGE, YELLOW, GREEN, BLUE, AND VIOLET, SEEN WHEN SUNLIGHT PASSES THROUGH A PRISM.

Effect on the Moon

The way we see the moon each night is completely dependent on the sun's rays. The moon shines because sunlight bounces off it. How much of the moon we see depends on its position in comparison to the sun and Earth. A new moon occurs when the moon is between the sun and Earth. A full moon happens when Earth is between the sun and moon.

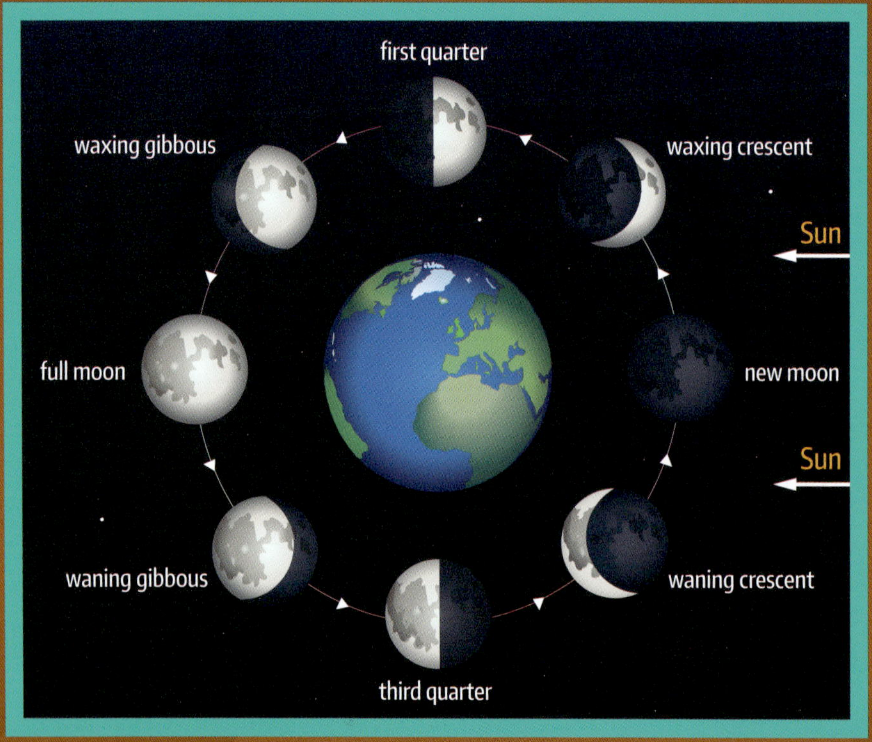

The moon repeats its phases, or shapes, once every 29.5 days as it orbits the Earth. Each phase has its own name.

The moon is orange during a lunar eclipse from the way the sun's rays bounce off Earth's atmosphere.

During the new moon, a solar eclipse can occur. Passing in front of the sun, the moon appears as a black circle. What looks like a ring of fire appears around the circle. This ring is the sun's corona. If Earth completely blocks the sun, a lunar eclipse can occur. The moon is then in Earth's shadow. The moon glows a dim red orange because some sunlight reaches it indirectly. Many cultures on Earth have stories or myths to explain why the eclipses happen.

Consider This

If you had no scientific knowledge of the sun or moon, what story might you make up to explain what they are?

Numbered Days

The sun is 4.6 billion years old. It formed from a rotating cloud of dust known as a **nebula**. Gravity caused most of the dust to be pulled toward the center, where it formed the sun. The remaining dust formed the planets.

This drawing shows the life cycle of our sun and what it will look like in the future.

The sun has been a constant ball of energy since its formation. But in another few billion years, the sun will expand into a red giant star. Then it will send out part of its mass into space. What remains of the sun will be known as a white dwarf star. It will continue to shine for billions of years, even as it gradually cools.

That won't happen for a long, long time. In the meantime, living things will continue to live and grow because of the sun's light and heat.

WORD WISE

A NEBULA IS A GIANT CLOUD OF GAS OR DUST IN SPACE.

Glossary

axis A straight line through a body, such as the sun or Earth.
classify The act of separating things into categories based on shared characteristics.
constellation A grouping of stars as viewed from Earth.
energy The power to work or to act.
galaxy One of the very large groups of stars and other matter that are found throughout the universe.
gravity A force of attraction between bodies that occurs because of their masses.
orbit The path taken by one body circling around another body.
particle One of the very small parts of matter (as a molecule, atom, or electron).
pressure The force exerted over an area by a substance.
solar eclipse The total or partial hiding of the sun by the moon's shadow.
solar system A star with the group of heavenly bodies that revolve around it, such as planets, moons, asteroids, and comets.
solstice The two times each year when the sun is at its greatest distance from the celestial equator.
tilt To incline at an angle.
universe All of space and everything in it including stars, planets, and galaxies.

For More Information

Books

Cock-Starkey, Claire. *Lore of the Stars: Folklore & Wisdom from the Skies Above.* Beverly, MA: Wide Eyed Editions, 2023.

Schuh, Mari C. *The Sun: The Center of Our Solar System.* Minneapolis, MN: Jump!, 2023.

Thorpe, Judy. *The Sun.* Buffalo, NY: Cavendish Square, 2024.

Websites

Britannica Kids: Sun
www.kids.britannica.com/kids/article/Sun/353824
Diagrams help you understand more about how the sun provides energy to Earth.

NASA Space Place: All About the Sun
www.spaceplace.nasa.gov/all-about-the-sun/en
Read more about the sun and look at stunning pictures on this NASA website for kids.

Index